Quick Reference of Possible

SIGNS & SYMPTOMS*

	NERVE AGENTS	BLISTER AGENTS		CHOKING AGENTS	
	GB (Sarin) GD (Soman) GA (Tabun) GF and VX	Sulfur Mustard H HD HS	Lewisite L HL (Lewisite-Mustard mixture)	Phosgene	Chlorine
MILD SYMPTOMS					
runny nose	■				
chest tightness	■			■	
shortness of breath	■				
rhinorrhea (excessive mucus from nose)		■	■		
sneezing		■	■		
cough – choking sensation				■	■
nosebleed		■			
hoarseness		■			
hacking cough		■			
eye pain / burning	■	■	■	■	■
eye itching & tearing		■		■	
eyes – gritty feeling		■			
pinpoint pupils	■				
bleeding in eyes				■	
erythema (skin redness)		■		■	
immediate burning pain at sites of exposure			■		
pain of mouth, nose, pharynx				■	
headache	■			■	
behavioral changes	■				
MODERATE SYMPTOMS – all mild symptoms, plus the following:					
increasing chest tightness / difficulty breathing	■			■	
cough & frothy mucous secretions			■		
increased eye symptoms / blurred vision	■				
eye reddening & irritation		■	■		
eye lid edema		■	■		
abdominal cramps	■				
diarrhea	■		■		
nausea	■			■	

...ible illness due to chemical agents. It is not meant to replace the expertise or professional advice of a physician or trained healthcare provider."

Quick Reference of Possible

SIGNS & SYMPTOMS*

	NERVE AGENTS	BLISTER AGENTS		CHOKING AGENTS	
	GB (Sarin) GD (Soman) GA (Tabun) GF and VX	Sulfur Mustard H HD HS	Lewisite L HL (Lewisite-Mustard mixture)	Phosgene	Chlorine
MODERATE SYMPTOMS continued					
vomiting				■	
headache, confusion, drowsiness	■				
weakness	■		■		
twitching of large muscle groups	■				
drooling and excessive sweating	■				
SEVERE SYMPTOMS – all mild and moderate symptoms, plus the following:					
respiratory failure	■				
burning sensation in chest					■
rapid breathing / cyanosis					■
productive cough		■			
mild to severe dyspnea		■			
permanent eye damage or blindness			■		
eye lid edema – severe		■			
possible corneal damage		■			
eye pain – severe		■			
erythema (skin redness) & blistering		■	■		
progressive increase in pain at sites of exposure			■		
progressive development of severe lung edema			■	■	■
subnormal temperature			■		
low blood pressure			■		
life-threatening burns			■		
shock				■	
involuntary defecation / urination	■				
very copious secretions	■				
twitching, jerking	■				
seizures	■				
flaccid paralysis	■				
coma	■				

* "This list of Signs & Symptoms is intended to assist in the assessment of possible illness due to chemical agents. It is not meant to replace the expertise or professional advice of a physician or [illegible]"

About this Guide

In the event of a terrorist attack, the Internet and other electronic resources may not be available. This guide was developed for first responders and caregivers at the local, State, and Federal levels to provide quick, concise, on-scene access to information about the most likely chemical/nuclear weapons. Read through this guide to become familiar with important elements of response including personal protection, casualty management, decontamination, and information about specific agents. Keep it with you in the field for quick reference in the event of a chemical/nuclear weapon attack. It is hoped that the use of this book is never required.

Advisory System (established March 2002)

Low (green)

Low risk of terrorist attacks

- Refine and exercise, as appropriate, preplanned protective measures;
- Ensure personnel receive proper training on the Homeland Security Advisory System and specific preplanned department or agency protective measures; and
- Institutionalize a process to assure that all facilities and regulated sectors are regularly assessed for vulnerabilities to terrorist attacks, and all reasonable measures are taken to mitigate these vulnerabilities.

Guarded (blue)

General risk of terrorist attacks

- Check communications with designated emergency response or command locations;
- Review and update emergency response procedures; and
- Provide the public with any information that would strengthen its ability to act appropriately.

Elevated (yellow)

Significant risk of terrorist attacks

- Increase surveillance of critical locations;
- Coordinate emergency plans as appropriate with nearby jurisdictions;
- Assess whether the precise characteristics of the threat require the further refinement of preplanned protective measures; and
- Implement, as appropriate, contingency and emergency response plans.

High (orange)

High risk of terrorist attacks

- Coordinate necessary security efforts with Federal, State, and local law enforcement agencies or any National Guard or other appropriate armed forces organizations;
- Take additional precautions at public events and possibly consider alternative venues or even cancellation;
- Prepare to execute contingency procedures, such as moving to an alternate site or dispersing workforce; and
- Restrict threatened facility access to essential personnel only.

Severe (red)

Severe risk of terrorist attacks

- Increase or redirect personnel to address critical emergency needs;
- Assign emergency response personnel and pre-position and mobilize specially trained teams or resources;
- Monitor, redirect, or constrain transportation systems; and
- Close public and government facilities.

General Preparedness

General Preparedness

Definition of Chemical Weapons

Toxic chemicals, ammunition and their dispersal equipment which through their chemical effects on living processes may cause death, temporary loss of performance or permanent injury to people and animals.

Classification of Chemical Weapon Agents

- Lethal **OR** Incapacitating (generally a fraction of the lethal dose)

AND

- Non-persistent, volatile, contaminate air, evaporate quickly **OR**
- Persistent, non-volatile, oily, viscous or thickened, will remain in the environment, will complicate decontamination

Controlling the Scene

Assessment of hazard & exposure potential

Things to consider:

- Method of dissemination of agent
- Size of affected area
- Environmental conditions – agent released indoors or outdoors, wind speed, direction, etc.
- Number of people in immediate area

Assessment of personal protection needs

- Level A equipment is required
- Determine isolation and evacuation distances
- Evacuate unaffected individuals upwind from incident site according to state/local guidelines
- **Shelter in place** if evacuation not possible or practical

Personal Protection

RESPONDING TO AN EVENT

LEVEL A Personal Protective Equipment (PPE)

Provides highest level protection for respiratory, skin, eye, and mucous membrane danger

- Respiratory protection
  - NIOSH-approved positive-pressure, self-contained breathing apparatus (SCBA) – replace canisters and cartridges before using respirator
- Chemical protective clothing
  - Fully encapsulating chemical vapor protective suit
  - Use a Toxic Agent Protective Apron (TAP Apron) for decontamination activities and patient handling
  - Chemical goggles and face shield
  - Chemical resistant double gloves – replace frequently
  - Chemical protective boots
- Communication
  - Use two-way radio system

NOTE When responding to an unknown agent, use the highest level of personal protection available. Assume that the agent is highly toxic and potentially lethal.

NIOSH – National Institute of Occupational Safety & Health

General Preparedness continued

REMOVAL OF PERSONAL PROTECTION

1. Hi-pressure, low-volume shower (minimum x 2) for responders in Level A suits
2. Wash down PPE with 5% hypochlorite solution (if available)
3. After complete decontamination, remove protective suit and shoe covers
4. Remove gloves and wash hands with soap and water
5. Remove hood, mask, and eye protection
6. Shower and wash with soap and water or wash down with 0.5% hypochlorite solution

Isolation & Evacuation Distances

ISOLATE & EVACUATE THE AREA DOWNWIND FROM INCIDENT SITE

Isolation Distances		Small Incident	Large Incident
Nerve Agents		700	2000
Blister Agents	(feet)	700	2000
Choking Agents		400–1100	400–1100

Evacuation Distances		Small Incident		Large Incident	
		Day	Night	Day	Night
Nerve Agents		1.2	5.5	4.6	7.0
Blister Agents	(miles)	1.2	5.5	4.6	7.0
Choking Agents		0.5	1.7	1.5	6.0

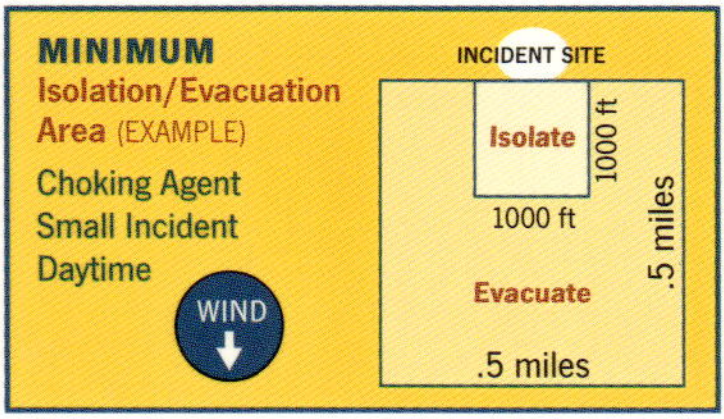

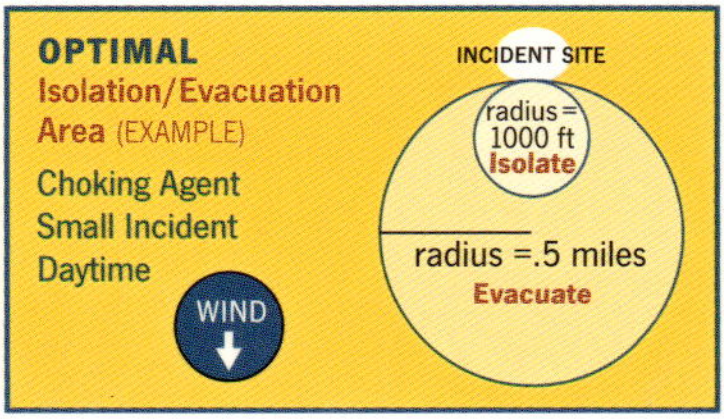

Emergency Actions

Control the Scene

- Approach from upwind with caution
- Determine safe site entry location
- Establish communications
- Ensure safety of personnel
    - Thorough hazard assessment
    - Use PPE
    - Use buddy system
    - Monitor and respond to changing conditions
- Contain/neutralize release
- Establish Site Control Zone (see chart at right)
    - Establish incedent control line
    - Limit access to authorized personnel
    - Establish outer control line (optional)
    - Reroute traffic and establish strict traffic control corridor into and out of contaminated area
    - Evacuate or shelter in place as indicated (see page 19)

General Preparedness continued

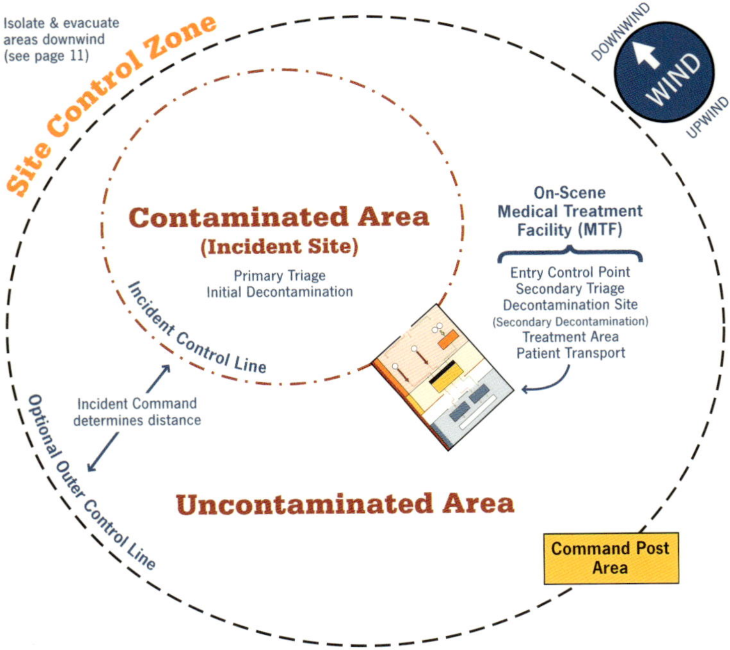

- Establish on-scene Medical Treatment Facility (see pgs 34 to 39)
  - Hot Zone – secondary triage, clothing removal
  - Warm Zone – Decontamination Site
  - Cold Zone – treatment and transport areas
- Establish and maintain crowd control
  - Prevent panic
  - Prevent public disorder

Emergency Actions continued

Rescue Victims

- Triage victims (see Triage section)
- Provide basic life support (ABC's)
- Evacuate victims/personnel to on-scene MTF
- Secondary triage for yellow and green tag patients
- Disrobe victims – clothing removal is a major part of decontamination
- Decontaminate victims (see Decontamination section)
- Administer antidotes, if required
- Provide Advanced Trauma Life Support (ATLS)
- Notify treatment facilities, local, state agencies
- Transport victims (see Patient Transport on page 17)

Additional Response Actions

- Evaluate rescuers and equipment post-entry
- Conduct thorough site evaluation
- Preserve incident site as crime scene
- Collect evidence
- Establish/conduct proper record keeping
  - Names and addresses of all persons involved
  - Make detailed records of the incident
- Accomplish site isolation and clean-up
- Remove protective gear/clothing at incident control line (see page 11 for removal of personal protection)

Initial Response

Initial Response

Pre-Hospital Response

- Coordinate local response with regional/state response plan
- Notify receiving hospitals and appropriate agencies
- Check Personal Protective Equipment (PPE)
- Review protocols for site control, triage, decontamination, casualty management, and transport

Patient Triage

Goal: provide maximum benefit to greatest number of victims

- Responder most experienced in patient evaluation should conduct triage
- Triage decisions based upon:
    - Local resources
    - Degree of patient contamination
    - Type of chemical weapons agent
    - Patient's clinical status
    - Additional injuries

Patient Transport

- Coordinate with overall response plan
- Decontaminate patients prior to transport
- Transport to designated hospital
- No transport without prior notification
    - Number of patients
    - Types and severity of injuries
- Use ground transport, NOT AIR
- Identify decontaminated patients clearly

Initial Response continued

Hospital Emergency Management

- Review protocols for evaluation, decontamination and treatment
- Determine agent identity
- Notify pharmacy if nerve agent is involved
- Breakout PPE, decontamination supplies (soap, water, hypochlorite solution), antidotes, etc.
- Perform pre-arrival determinations
  - Number of victims
  - Medical condition of individual victims
  - Treatment rendered to victims/decontamination status
  - Estimated Time of Arrival (ETA)
- Notify local emergency authorities
- Establish security & traffic control of hospital entrances & grounds
- Clear and secure all areas to prevent contamination
- Establish methods for contamination control
- Prevent entry of victims into emergency department or MTF without adequate evaluation and decontamination
- Establish controlled area for patient arrival, decontamination, treatment and holding (Hot, Warm, Vapor Hazard & Cold Zones)
- Establish control line between controlled & uncontrolled areas
- Secure and control traffic to and from controlled area
- Establish secondary control line
- Prevent tracking of contamination by covering floor areas (brown wrapping paper, butcher paper, cloth sheets, taped securely to floor)
- Establish strict isolation precautions
- Control waste disposal and ventilation
- Decontaminate patients before transfer out of controlled area

Initial Response continued

- Verify patient and personnel decontamination prior to exit from controlled area
- Transfer patients only with prior notification
- Establish accurate record keeping for patient management and identification

NOTE: Hospitals should be aware that contaminated patients may arrive at treatment facilities without prior notification

Contained Casualty Setting (relatively few victims)

- Decontaminate patient outdoors
- Capture and dispose of runoff

Mass Casualty Setting (large number of victims)

- Decontaminate patient outdoors
- Capture runoff, if possible
  (environmental concerns are secondary to saving lives)

Sheltering in Place

- When evacuation is not possible or practical
- People are instructed to stay where they are – home, work, school, etc.
- Go to a room with as few doors and windows as possible
- Chemical gases sink – choose highest level of building
- If possible, seal room with plastic or duct tape
- Listen to the radio for instructions from emergency coordinators

Triage

Triage

Goal: provide maximum benefit to greatest number of victims

- Responder most experienced in patient evaluation should conduct triage
- Triage decisions based upon:
    - Local resources
    - Degree of patient contamination
    - Type of chemical weapons agent
    - Patient's clinical status
    - Additional injuries

KEY TERMS

Ambulatory
Able to understand directions, talk, and walk unassisted

Non-ambulatory
Unconscious, unresponsive, or unable to move unassisted

Considerations for high priority triage:

- Victims closest to point of release
- Victims reporting exposure to vapor or aerosol
- Victims with evidence of liquid on skin or clothing
- Victims with serious medical symptoms (shortness of breath, chest tightness, etc.)

Triage Procedures

1. Patient alive? . . . . IF **NO** 

2. Breathing compromised?. . . . . IF **YES** 

3. Circulatory system intact?. . . . . IF **NO** 

Divide patients into ambulatory and non-ambulatory groups and continue for both groups

4. Serious signs or symptons of medical distress or chemical exposure? . . . . IF **YES** 

5. Moderate signs or symptoms of liquid exposure or close proximity to release point? . . . IF **YES** 

6. Minimal signs or symptoms, vapor exposure, or close proximity to release point?. . . . . IF **YES** 

7. Minimal signs or symptoms with no exposure to vapor or liquid? . . . . IF **YES** 

Triage continued

S.T.A.R.T Category	Actions	Prioritization
IMMEDIATE Red Tag **Category 1**	**Immediate** **1.** Decontamination **2.** Interventions **3.** Treatment	**1.** Breathing compromised **2.** Non-Ambulatory **3.** Ambulatory
DELAYED Yellow Tag **Category 2**	**Delayed** **1.** Decontamination **2.** Treatment	**1.** Non-Ambulatory **2.** Ambulatory
MINOR Green Tag **Category 3**	**Low priority** **1.** Decontamination **2.** Treatment	**1.** Non-Ambulatory **2.** Ambulatory
DECEASED/ EXPECTANT Black Tag **Category 4**	**Lowest priority** **1.** Decontamination	

S.T.A.R.T. – Simple Triage And Rapid Transport

Decontamination

Decontamination

Principles of Decontamination

- Method of decontamination is not as important as how and when it is used
- Remove agent by best means available
- Decontamination
    - Soap, water, and hypochlorite solutions are very effective
    - 0.5% hypochlorite solution with an alkaline pH (10) is best available UNIVERSAL liquid decontamination agent
- Vapor and aerosols may exist in environment after chemical attack
- Vapors may be trapped inside clothes
- Initial incident site decontamination must be followed by a complete secondary decontamination at on-scene Medical Treatment Facility

Purpose of Decontamination

- Removal of agent from skin and clothing
- Reduction of further exposure to agent of victims and responders
- Maintenance of uncontaminated treatment facility

The first 1 to 2 minutes following ANY chemical exposure is the most important and most effective time for decontamination

Initial Incident Site Decontamination

- Airway management may be required prior to decontamination (severely intoxicated nerve agent patients)
- Antidote may need to be provided emergently if nerve agent
- Physical removal of agent

  (works well on chemical agents regardless of chemical structure)
  - Scrape or blot off visible agent from skin
  - Disrobe victims head to toe
  - Do not pull contaminated clothing over head and face, cut it off
  - Flushing with large quantities of water
  - Absorbents to soak up the agent
    - Military decontamination kits (M291 and M295) commercially available – suitable for small areas
    - RSDL neutralizer (military use only)
    - Other: dirt, flour, talc, vermiculite, soap detergents

Secondary Decontamination

Contained Casualty Setting (relatively few victims)

Ambulatory Patients

- Physical removal of agent
    - Scrape or blot off visible agent from skin
    - Disrobe victims head to toe
    - Do not pull contaminated clothing over head and face, cut it off
- Decontamination
    - Flush or shower with large quantities of water for 2 to 3 minutes is ideal
        - High-volume, low-pressure (60 to 90 psi) shower
    - If available
        - Use 0.5% hypochlorite solution to wash skin
        - Use soap and water to wash skin
- Wound Decontamination
    - Open wounds may be contaminated by chemical agents (nerve agents, vesicants)
    - Wound irrigation with 0.5% hypochlorite solution may be effective in conjunction with wound debridement
    - Thickened agents are difficult to remove from wounds and often fatal
    - Fabric and foreign bodies in wounds may contain and slowly release chemical agent
    - Risk of vapor off gassing from material in wounds is very low

Secondary Decontamination continued

Contained Casualty Setting (relatively few victims)

Ambulatory Patients continued

- Eye care
  - Remove contact lenses, eyeglasses
  - Flush eyes with water or saline solution for several minutes
- Redress in uncontaminated clothing
- Contaminated clothing should be sealed inside a plastic bag, and then sealed inside a second plastic bag. Follow EPA guidelines for disposal.

Non-Ambulatory Patients

- Emergency personnel
  - Dress in Level A PPE and TAP Apron
  - 2 people per litter
    - Work with one patient on litter from triage to "hand-off" at "Hot Line," the point between contaminated and agent-free zone
  - 2 additional assistants for clothing removal/decontamination
- Decontamination
  - Warm soap and water wash x 3 **OR**
  - Hypochlorite Solutions
    - 0.5% hypochlorite solution for patient washing
    - 5% hypochlorite solution for equipment/litter decontamination
  - Keep solutions in separate & clearly marked buckets

Secondary Decontamination continued

Mass Casualty Setting (large number of victims)

- Water flushing is the BEST mass casualty decontamination method
- Showering time may be shorter when numbers of victims requiring decontamination are great

First Responder Decontamination Procedures

1. Hi-pressure, low-volume shower (minimum x 2) for responders in Level A suits
2. Wash down PPE with 5% hypochlorite solution (if available)
3. After complete decontamination, remove protective suit and shoe covers
4. Remove gloves and wash hands with soap and water
5. Remove hood, mask, and eye protection
6. Shower and wash with soap and water or wash down with 0.5% hypochlorite solution

- Follow NFPA 471 "Recommended Practices for Response to Hazardous Materials Incidents"

On-Scene Medical Treatment Facility (MTF)

- Components of an on-scene MTF:
  - Hot Zone – contaminated receiving area, secondary triage, clothing removal
  - Warm Zone – Decontamination Site (DS)
  - Vapor Hazard Zone – barrier between contaminated and agent-free zones
  - Cold Zone – treatment and transport area
- Can be a 1 or 2 corridor system (see pages 36 to 39)

HOT ZONE

- Personnel must wear Level A PPE

WARM ZONE

- Personnel must wear Level A PPE

MEDICAL TREATMENT FACILITY – KEY ELEMENTS

- Determine wind direction
  - MTF upwind from incident site
  - Adaptable to rapid adjustment if wind direction changes
  - Wind change of any direction requires evaluation and possible adjustment
- Establish security and strict traffic control
- Clear and secure all areas to prevent contamination
- Establish Medical Treatment Facility
  - Entry Control Point (ECP)
  - Hot Zone
  - Warm Zone – Decontamination Site
  - "Hot Line"
    - Demarcation between Warm Zone and Vapor Hazard Zone
    - Strictly controlled access points
    - Engineer tape or concertina wire to prevent uncontrolled access
  - Vapor Hazard Zone – wide enough to minimize liguid and vapor hazard (25 to 50 yards)
  - Cold Zone
- Establish strict isolation precautions
- Control waste disposal and ventilation

Emergency Decontamination Corridor Systems

1 Corridor System

Use 1 corridor system if limited personnel

1. Triaged victims enter MTF through Hot Zone
2. Red tag patients are first priority for decontamination and treatment
3. Yellow and green tag patients are separated into ambulatory and non-ambulatory groups
4. Patients' clothing is removed before entering Decontamination Site (Warm Zone)
5. Decontaminated patients enter Cold Zone for treatment and/or transport to hospital

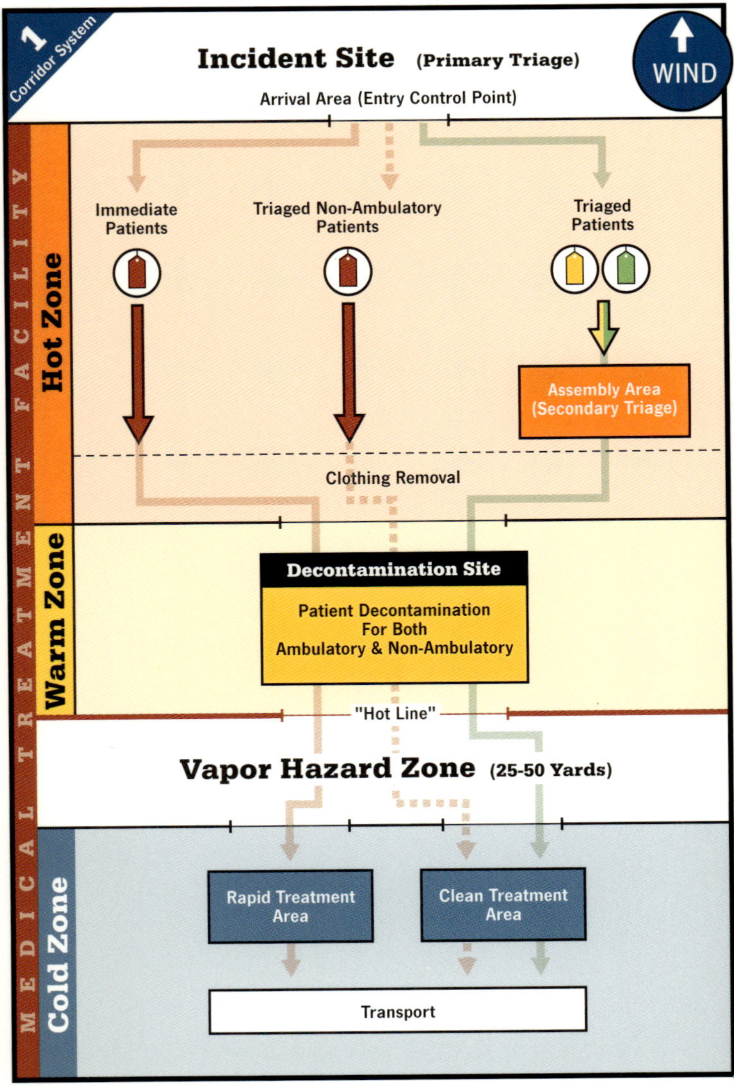
1 Corridor System
Incident Site (Primary Triage)
WIND
Arrival Area (Entry Control Point)
MEDICAL TREATMENT FACILITY
Hot Zone
Immediate Patients
Triaged Non-Ambulatory Patients
Triaged Patients
Assembly Area (Secondary Triage)
Clothing Removal
Warm Zone
Decontamination Site
Patient Decontamination For Both Ambulatory & Non-Ambulatory
"Hot Line"
Vapor Hazard Zone (25-50 Yards)
Cold Zone
Rapid Treatment Area
Clean Treatment Area
Transport

Emergency Decontamination Corridor Systems

2 Corridor System

Use 2 corridor system if enough personnel

1. Triaged victims enter MTF through Hot Zone
2. Red tag patients are first priority for decontamination and treatment
3. Yellow and green tag patients are separated into ambulatory and non-ambulatory groups
4. Patients' clothing is removed before entering Decontamination Site (Warm Zone)
5. Decontamination Site has separate areas for ambulatory and non-ambulatory patients
6. Decontaminated patients enter Cold Zone for treatment and/or transport to hospital

NOTE Depending on local protocols, responders in the Hot Zone may perform some treatments (antidote injections, ABC's)

Decontamination continued

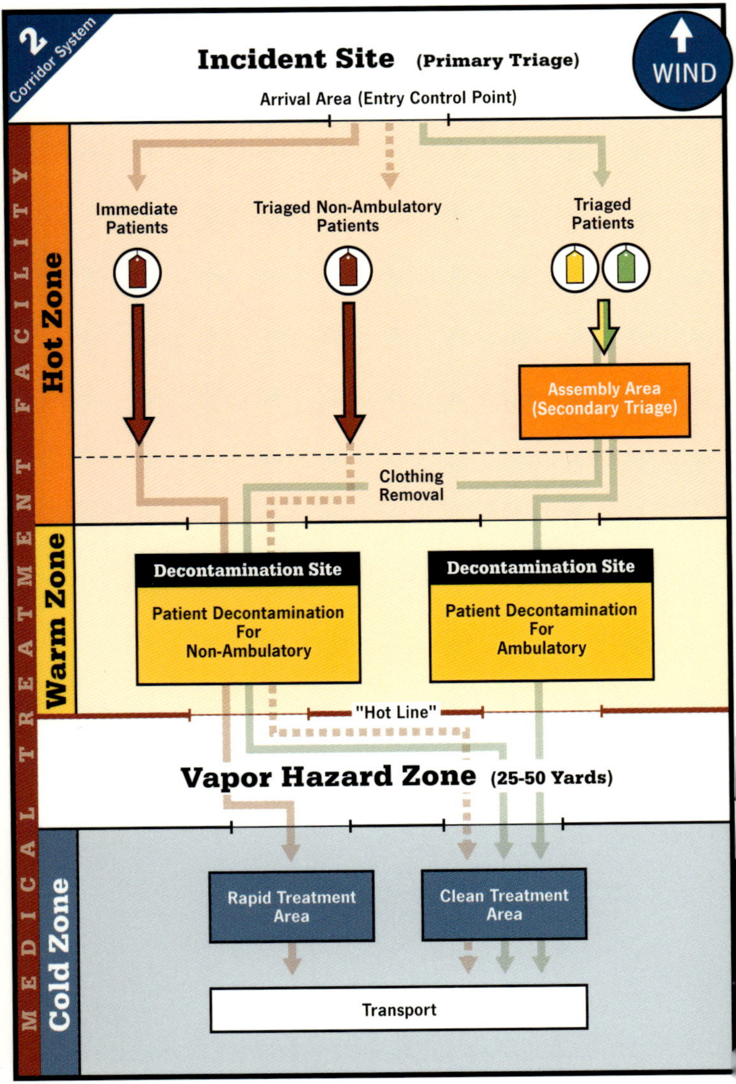

Other Types of Mass Decontamination Systems

- Ladder type (see diagrams on next two pages)
- Commercial decontamination systems
- Other methods – may be used when other systems not available
    - Overhead fire sprinkler systems
    - Public fountains
    - Chlorinated swimming pools
    - Dry gel or powdered decontamination agents (dirt, flour, Fuller's earth, baking powder, sawdust, charcoal, ashes, activated carbon, alumina, silica gels, zeolites, clay materials, and tetracalcium aluminate)
    - Reactive foams

Environmental and Legal Concerns

- EPA gives highest priority to responder actions to save lives and preserve health during chemical terrorist incident
- Once imminent threat to life and health addressed, first responders should take all reasonable efforts to contain or mitigate environmental consequences
- Involve state and federal officials ASAP to reduce potential liability
- Obtain legal counsel in the individual state to discuss authority, status as an agent of the state, immunities and indemnification

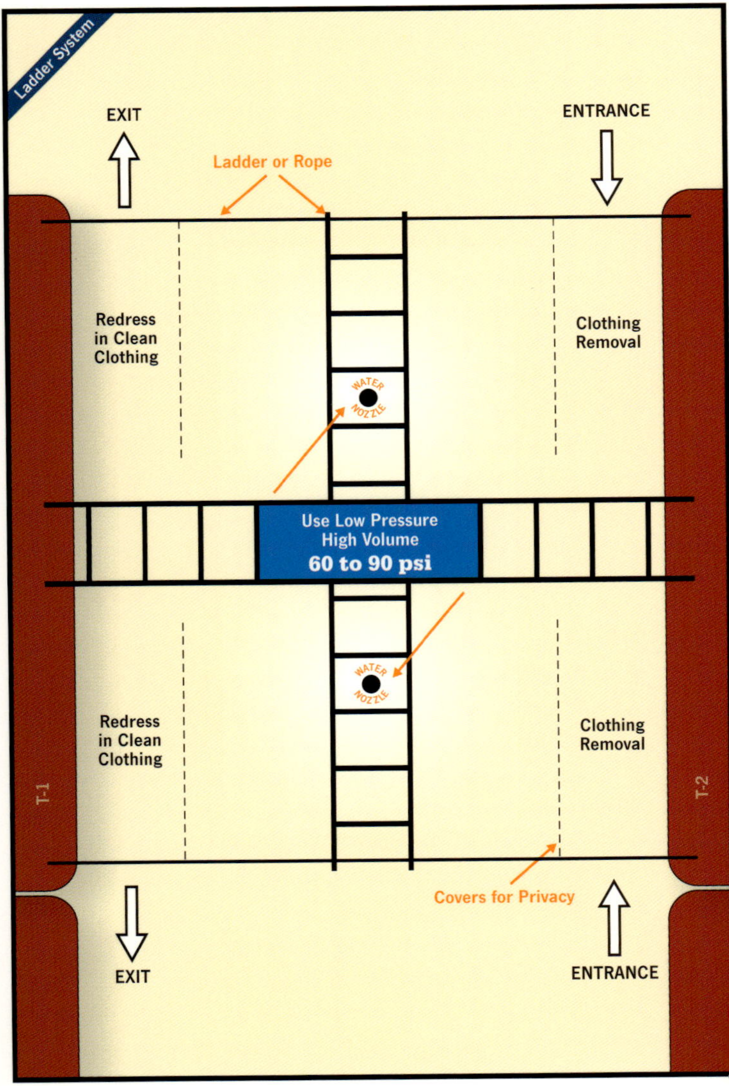
Ladder System
EXIT
ENTRANCE
Ladder or Rope
Redress in Clean Clothing
Clothing Removal
WATER NOZZLE
Use Low Pressure
High Volume
60 to 90 psi
WATER NOZZLE
Redress in Clean Clothing
Clothing Removal
T-1
T-2
Covers for Privacy
EXIT
ENTRANCE

Decontamination continued

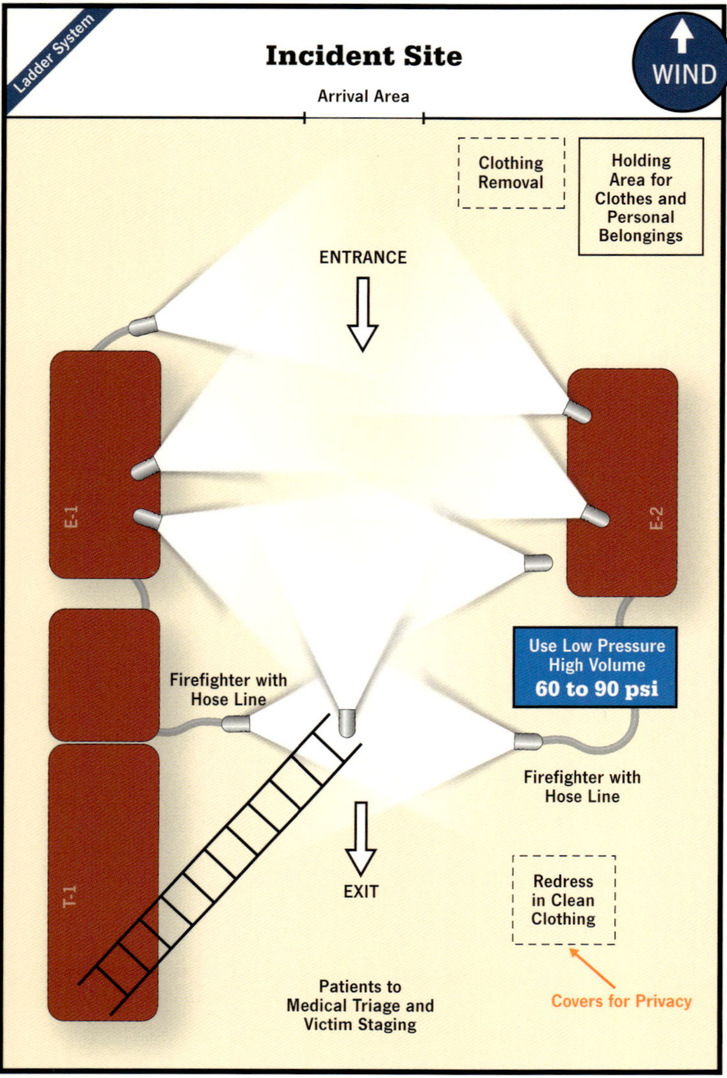

Nerve Agents

Nerve Agents

Contact Risk

- **High** – avoid contact with liquid and vapor

Agents

GA (Tabun)

GB (Sarin)

GD (Soman)

GF

VX

Antidote

Atropine and 2-PAM Cl (pralidoxime chloride)

Routes of Exposure

- **Inhalation**
- **Ingestion**
- **Skin/Eye contact**

Personal Protection

- **See General Preparedness section for specific Personal Protection advice**

Emergency Actions

- **See General Preparedness section for specific Emergency Action advice**

NOTES

- Rescuers can be contaminated by direct contact with the clothing or skin of victims or by vapors trapped in their clothing
- Clothing can trap and release agent for about 30 minutes after contact with vapor

Nerve Agents continued

Signs & Symptoms

Initial effects may occur within seconds (loss of consciousness and convulsions) or hours depending on dose and route of exposure:

VAPOR = seconds to minutes

LIQUID = minutes to hours

MILD

- Runny nose
- Chest tightness
- Pinpoint pupils
- Shortness of breath
- Eye pain
- Headache
- Behavioral changes

MODERATE – all mild symptoms, plus the following:

- Increased eye symptoms/blurred vision
- Drooling and excessive sweating
- Increasing chest tightness/difficulty breathing
- Nausea
- Vomiting
- Abdominal cramps
- Diarrhea
- Weakness
- Twitching of large muscle groups
- Headache, confusion, drowsiness

Nerve Agents continued

SEVERE – **all mild and moderate symptoms, plus the following:**

- Involuntary defecation/urination
- Very copious secretions
- Twitching, jerking
- Seizures
- Flaccid paralysis
- Coma
- Respiratory failure
- Death

Characteristics

Most Toxic Chemical Warfare Agents

- Variable volatility
- Liquids at room temperature
- Chemically similar to organophosphate pesticides
- Inhibit acetylcholinesterase enzymes
- Vapors are heavier than air
- All cause same signs and symptoms by inhibiting acetylcholinesterase enzymes thus disrupting nerve impulse transmission
- May produce rapidly fatal systemic effects

Characteristics continued

"G" Agents

- Generally non-persistent
- May be thickened to increase viscosity and persistence
- Clear, colorless, tasteless liquids
- GA and GD have slight odors

VX

- Persistent: will remain on ground several weeks (longer in colder climates)
- Clear, oily, amber-colored liquid
- Odorless

Treatment: Pre-Hospital

HOT ZONE

(Use Level A PPE)

- Execute ABC's on victims
- Triage patients (see Triage section)
- Check for obstructed airway – clear if necessary
- Administer antidotes, if available and symptoms severe
- Move victims to Decontamination Site – for non-ambulatory, use gurney or backboard that can be decontaminated

NOTE Maintain records of all medications given

WARM ZONE (Decontamination Site)

(Use Level A PPE)

- Confirm clear airway
- Administer antidotes, if necessary
- Perform decontamination (see Decontamination section)
- No transfer to Cold Zone until completely decontaminated

COLD ZONE

- Verify decontamination
- Execute ABC's
- Administer antidotes, if necessary
- Perform advanced treatment as necessary
    - Advanced life support protocols: coma, hypotension, cardiac arrythmias, and seizures
    - Treat convulsions: Diazepam – Adults 5 to 10 mg, Children 0.2 to 0.5 mg/kg
    - Treat 2-PAM Cl induced hypertension: Phentolamine – Adults 5 mg IV, Children 1 mg IV
- Transport to hospital
    - Contact and give them condition of patient, treatment already given, and Estimated Time of Arrival (ETA)

Nerve Agents continued

Treatment: Pre-Hospital Guidelines

Patient Age	Mild/Moderate Symptoms	Severe Symptoms
INFANT (0 to 2 yrs)	**Atropine** 0.05 mg/kg IM **2-PAM Cl** 15 mg/kg IM	**Atropine** 0.1 mg/kg IM **2-PAM Cl** 25 mg/kg IM
CHILD (2 to 10 yrs)	**Atropine** 1 mg IM **2-PAM Cl** 15 mg/kg IM	**Atropine** 2 mg IM **2-PAM Cl** 25 mg/kg IM
ADOLESCENT (over 10 yrs)	**Atropine** 2 mg IM **2-PAM Cl** 15 mg/kg IM	**Atropine** 4 mg IM **2-PAM Cl** 25 mg/kg IM
ADULT	**Atropine** 2 to 4 mg IM **2-PAM Cl** 600 mg IM	**Atropine** 6 mg IM **2-PAM Cl** 1800 mg IM
ELDERLY, FRAIL	**Atropine** 1 mg IM **2-PAM Cl** 10 mg/kg IM	**Atropine** 2 to 4 mg IM **2-PAM Cl** 25 mg/kg IM

Treatment: Pre-Hospital Guidelines continued

Other Treatments	
Assisted ventilation for severe exposures after administering antidotes	**Repeat Atropine (2 mg IM)** at 5 to 10 minute intervals until secretions have diminished and breathing is comfortable or airway resistance is at or near normal

Treatment: Emergency Department

NOTE Contaminated patients may arrive at ER without notification

DECONTAMINATION AREA

(Use Level A PPE for decontamination)

- Decontaminate prior to entry into facility
- Only decontaminate inside hospital if:
  - Use negative air pressure room
  - Use floor drains to contain contamination
- Execute ABC's

TREATMENT AREA

- Verify decontamination
- Triage patients (see Triage section)
- Observe for at least 18 hours – patients with exposure to liquid agent – even if no immediate symptoms

NOTE Effects of skin exposure may take up to 18 hrs to develop

Nerve Agents continued

Treatment: Emergency Department continued

- Discharge – patients with exposure to vapor only and no symptoms
- Admit/Treat – patients with miosis/rhinorrhea and eye pain, headache, vomiting
  - Treat miosis: topical Atropine or Homatropine
  - Treat severe rhinorrea: Atropine IM – Adults 2 mg, Children 0.05 mg/kg
  - Treat mild/moderate, and severe symptoms (see below)

Treatment: Emergency Department Guidelines

Patient Age	Mild/Moderate Symptoms	Severe Symptoms
INFANT (0 to 2 yrs)	**Atropine** 0.05 mg/kg IM **OR** 0.02 mg/kg IV **2-PAM Cl** 15 mg/kg IM slowly	**Atropine** 0.1 mg/kg IM **OR** 0.02 mg/kg IV **2-PAM Cl** 15 mg/kg IM slowly
CHILD (2 to 10 yrs)	**Atropine** 1 mg/kg IM **2-PAM Cl** 15 mg/kg IM slowly	**Atropine** 2 mg/kg IM **2-PAM Cl** 15 mg/kg IM slowly

Nerve Agents continued

Treatment: Emergency Depart. Guidelines continued

Patient Age	Mild/Moderate Symptoms	Severe Symptoms
ADOLESCENT (over 10 yrs)	**Atropine** 2 mg IM **2-PAM Cl** 15 mg/kg IV slowly	**Atropine** 4 mg IM **2-PAM Cl** 15 mg/kg IV slowly
ADULT	**Atropine** 2 to 4 mg IM **2-PAM Cl** 15 mg/kg (1 g) IV slowly	**Atropine** 6 mg IM **2-PAM Cl** 15 mg/kg (1 g) IV slowly
ELDERLY, FRAIL	**Atropine** 1 mg IM **2-PAM Cl** 5 to 10 mg/kg IV slowly	**Atropine** 2 mg IM **2-PAM Cl** 5 to 10 mg/kg IV slowly

Treatment: Emergency Depart. Guidelines continued

Other Treatments

Assisted ventilation as needed	**Repeat Atropine (2 mg IM)** at 5 to 10 minute intervals until secretions have diminished and breathing is comfortable or airway resistance is at or near normal
Diazepam for convulsions: **Infants to 5 years** 0.2 to 0.5 mg IV **Children over 5 years** 1 mg IV **Adults** 5 mg IV	**Phentolamine** for 2-PAM induced hypertension: **Children** 1 mg IV **Adults** 5 mg IV

Laboratory Tests/Findings

- CBC
- Glucose
- Serum electrolytes
- Chest x-ray
- Arterial blood gas measurements
- RBC (red blood cell) cholinesterase
- Plasma cholinesterase

Disposition/Follow-up/Reporting

- Monitor hospitalized patients closely
- Discharge patients with follow-up instructions
- After severe exposure follow-up for persistent CNS sequelae (central nervous system complications) up to 6 weeks after exposure
- Notify state or local health department of possible public health risk

Blister Agents #1

Sulfur Mustard Agents

H

HD

HS

Contact Risk

- **High** – avoid contact with liquid and vapor

Antidote

None

Routes of Exposure

- **Inhalation**
- **Ingestion**
- **Skin/Eye contact**

Personal Protection

- See General Preparedness section for specific Personal Protection advice

Emergency Actions

- See General Preparedness section for specific Emergency Action advice

NOTES

- Rescuers can be contaminated by direct contact with the clothing or skin of victims or by vapors trapped in their clothing
- Clothing can trap and release agent for about 30 minutes after contact with vapor

Blister Agents #1 continued

Signs and Symptoms

Progressive with delayed onset

NOTE The sooner after exposure that symptoms occur, the more likely they are to progress and become severe

Initial effects depend on dose, duration, and route of exposure:

- Clinical effects develop within 1 to 24 hours
- Agents produce skin, eye, and respiratory injury
- May cause bone marrow suppression, neurological and gastrointestinal toxicity
- Eyes are most sensitive to sulfur mustard agent

Clinical effects and time of onset by severity of exposure to Sulfur Mustard:

Tissue	Severity of Exposure	Clinical Signs and Symptoms	Time of Onset
EYES	Mild	Tearing, itching, burning, gritty feeling	4 to 12 hrs
	Moderate	Above effects and reddening, lid edema, moderate pain	3 to 6 hrs
	Severe	Marked lid edema, possible corneal damage, severe pain	1 to 2 hrs

Blister Agents #1 continued

Clinical effects and time of onset by severity of exposure to Sulfur Mustard: continued

Tissue	Severity of Exposure	Clinical Signs and Symptoms	Time of Onset
AIRWAYS	**Mild**	Rhinorrhea (excessive mucus from nose), sneezing, nosebleed, hoarseness, hacking cough	6 to 24 hrs
	Severe	Above effects and productive cough, mild to severe dyspnea	2 to 6 hrs
SKIN	**Mild**	Erythema (redness)	2 to 24 hrs
	Severe	Erythema (redness) and vesication (blistering)	2 to 24 hrs

Characteristics

- Yellow to dark brown/black in color
- Slight garlic or mustard odor
- Low volatility (evaporate slowly)
- Vapors are heavier than air
- Odor does not provide adequate warning of detection
- Clinical effects develop progressively over time

Blister Agents #1 continued

Treatment: Pre-Hospital

HOT ZONE

(Use Level A PPE)

- Execute ABC's on victims
- Triage patients (see Triage section)

Triage for Mustard Agent Casualties

Priority (Category)	Clinical Signs and Symptoms	Time of Onset
IMMEDIATE (Category 1)	Lower respiratory signs (dyspnea)	< 4 up to 12 hrs post exposure
DELAYED (Category 2)	Eye lesions with impaired vision; skin lesion covering 2 to 50% of body surface area for liquid exposure or any body surface burn for vapor exposure; lower respiratory symptoms (cough with sputum production, dyspnea)	> 4 hrs (eye & skin); OR > 12 hrs (respiratory) post exposure
MINIMAL (Category 3)	Minor eye lesion with no vision impairment; skin lesion < 2% of body surface area in noncritical areas; minor upper respiratory symptoms (cough, sore throat)	> 4 hrs post exposure
EXPECTANT (Category 4)	Lower respiratory signs (dyspnea); skin lesion covering 50% or more of body surface area from liquid exposure	> 4 hrs post exposure

Blister Agents #1 continued

- Check for obstructed airway – clear if necessary
- Move victims to Decontamination Site – for non-ambulatory, use gurney or backboard that can be decontaminated

NOTE Decontamination must be performed within 1 to 2 min following exposure to decrease tissue damage

NOTE Maintain records of all medications given

WARM ZONE (Decontamination Site)

(Use Level A PPE)

- Confirm clear airway – suction secretions and ventilate if necessary
- Provide oxygen if needed
- Perform decontamination (see Decontamination section)
- Ingestion – DO NOT induce vomiting
- No transfer to Cold Zone until completely decontaminated

COLD ZONE

- Verify decontamination
- Execute ABC's
- Secure airway – intubate as necessary
- Perform advanced treatment as necessary
    - Advanced life support protocols: coma, hypotension, cardiac arrythmias, and seizures
    - Treat bronchospasm with bronchodilaters
    - Intubation, cricothyrotomy

Blister Agents #1 continued

Treatment: Pre-Hospital continued

- Transport to hospital
  - Contact and give them condition of patient, treatment already given, and Estimated Time of Arrival (ETA)

NOTE Symptoms evolve over time – close observation and follow-up required

Treatment: Emergency Department

NOTE Contaminated patients may arrive at the ER without prior notification

DECONTAMINATION AREA

(Use Level A PPE for decontamination)

- Decontaminate prior to entry into facility
- Only decontaminate inside hospital if:
  - Use negative air pressure room
  - Use floor drains to contain contamination
- Execute ABC's

NOTE Patients arriving at the ER directly from scene (within 30 to 60 minutes) will rarely have symptoms

Blister Agents #1 continued

TREATMENT AREA

- Verify decontamination
- Triage patients (see Triage section)
- **Eye exposure**
    - Thorough eye exam
    - Visual acuity testing
    - Conjunctivitis
        - Mild symptoms – soothing eye drops
        - Moderate/severe symptoms – admit, observe, treat
- **Skin exposure**
    - Thorough evaluation
    - Small area – wound care, observe, discharge, follow-up
    - Large area – wound care, admit, observe, treat
- **Inhalation exposure**
    - Thorough evaluation
    - Mild symptoms after 12 hours post-exposure – treat symptoms and discharge with follow-up instructions
    - Moderate/severe symptoms – admit to critical care unit
- **Ingestion exposure**
    - DO NOT induce vomiting
    - Gastric lavage/suctioning for large ingestion

Blister Agents #1 continued

Laboratory Tests/Findings

- CBC
- Glucose
- Serum electrolytes
- Chest x-ray
- Arterial blood gases/pulse oximetry
- Urinalysis and thiodiglycol assay

Disposition/Follow-up/Reporting

- Admit patients with significant exposure
- Observe inpatients for delayed effects
- Instruct discharge patients to watch for delayed effects
- Follow-up laboratory evaluation for: pulmonary function and bone marrow toxicity
- Notify state or local health department of possible public health risk

Blister Agents #2

Contact Risk

- **High** – severe irritant, avoid contact with liquid and vapor

Lewisite Agents

L

HL (Lewisite-Mustard mixture)

Antidote

Dimercaprol (BAL – British Anti-Lewisite)

Routes of Exposure

- **Inhalation**
- **Ingestion**
- **Skin/Eye contact**

Personal Protection

- See General Preparedness section for specific Personal Protection advice

Emergency Actions

- See General Preparedness section for specific Emergency Action advice

NOTES

- Rescuers can be contaminated by direct contact with the clothing or skin of victims or by vapors trapped in their clothing
- Clothing can trap and release agent for about 30 minutes after contact with vapor

Blister Agents #2 continued

Signs and Symptoms

Rapid and progressive onset

Initial effects depend on dose, duration, and route of exposure:

- Immediate burning pain at sites of exposure
- Irritation and swelling of eyelids
- Permanent damage or blindness can occur in minutes
- Progressive increase in pain at sites of exposure
- Skin redness with blistering in 8 to 12 hours
- Profuse nasal secretions
- Violent sneezing
- Cough and frothy mucous secretions
- Progressive development of severe lung edema
- Restlessness and weakness
- Diarrhea
- Subnormal temperature
- Low blood pressure
- Life-threatening burns

Characteristics

- Amber to dark brown liquids
- Strong geranium or garlic-like odor
- Volatile
- Vapors are heavier than air
- Odors do not provide adequate warning of detection

Blister Agents #2 continued

- Produce rapid onset of symptoms
- Produce deeper and more severe burns than mustard agents
- Burns can be life-threatening
- Systemic absorption may occur following skin or eye exposure

Treatment: Pre-Hospital

HOT ZONE

(Use Level A PPE)

- Execute ABC's on victims
- Triage patients (see Triage section)

Triage for Lewisite Agent Casualties

Priority (Category)	Clinical Signs and Symptoms
IMMEDIATE (Category 1)	Lower respiratory signs (dyspnea, productive cough)
DELAYED (Category 2)	Eye lesions with impaired vision; moderate sized skin lesions for liquid exposure or any body surface burn for vapor exposure; lower respiratory symptoms (cough with sputum production)
MINIMAL (Category 3)	Minor eye lesion with no vision impairment; small skin lesions in non-critical areas; minor upper respiratory symptoms (cough, sore throat)
EXPECTANT (Category 4)	Lower respiratory signs (dyspnea, necrosis); skin lesion covering more than half of body surface area from liquid exposure

Blister Agents #2 continued

Treatment: Pre-Hospital continued

- Check for obstructed airway – clear if necessary
- Administer antidotes: Dimercaprol (BAL – British Anti-Lewisite)
    - Topical: ointment applied to exposed skin
    - Systemic: only for patients in shock or with significant pulmonary injury
    - BAL in oil: 3 to 6 mg, deep IM every 4 hours x 4 doses, every 2 hours in severe poisoning
- Move victims to Decontamination Site – for non-ambulatory, use gurney or backboard that can be decontaminated

NOTE Maintain records of all medications given

WARM ZONE (Decontamination Site)

(Use Level A PPE)

- Confirm clear airway
- Administer antidotes: Dimercaprol (BAL) British Anti-Lewisite
    - Topical: ointment applied to exposed skin
    - Systemic: only for patients in shock or with significant pulmonary injury
    - BAL in oil: 3 to 6 mg, deep IM every 4 hours x 4 doses, every 2 hours in severe poisoning
- Perform decontamination (see Decontamination section)
- Ingestion – DO NOT induce vomiting – give activated charcoal slurry if able to swallow
- No transfer to Cold Zone until completely decontaminated

Blister Agents #2 continued

COLD ZONE

- Verify decontamination
- Execute ABC's
- Secure airway – intubate as necessary
- Perform advanced treatment as necessary
  - Advanced life support protocols: coma, hypotension, cardiac arrythmias, and seizures
  - Treat bronchospasm with bronchodilaters
  - Intubation, cricothyrotomy
- Supportive care
- Transport to hospital
  - Contact and give them condition of patient, treatment already given, and Estimated Time of Arrival (ETA)

Treatment: Emergency Department

NOTE Contaminated patients may arrive at the ER without prior notification

DECONTAMINATION AREA

(Use Level A PPE for decontamination)

- Decontaminate prior to entry into facility
- Only decontaminate inside hospital if:
  - Use negative air pressure room
  - Use floor drains to contain contamination
- Execute ABC's

Blister Agents #2 continued

Treatment: Emergency Department continued

TREATMENT AREA

- Verify decontamination
- Triage patients (see Triage section)

NOTE Patients arriving at the ER directly from scene will have pain or irritation if they were exposed

- **Eye exposure**
  - Thorough eye exam
  - Visual acuity testing
  - Conjunctivitis
    - Mild symptoms – soothing eye drops
    - Moderate/severe symptoms/early onset – admit, observe, ophthalmology evaluation and treatment
- **Skin exposure**
  - Thorough evaluation
  - Small areas of erythema (redness) that begins after 12 hours – wound care, observe, discharge, follow-up
  - Large areas of involvement/early erythema (redness)/blistering – wound care, admit, observe, fluid resuscitation, treat (possibly in burn unit)
- **Inhalation Exposure**
  - Thorough evaluation
  - Mild symptoms after 12 hours post-exposure – treat symptoms and discharge with follow-up instructions
  - Moderate/severe symptoms – admit to critical care unit

Blister Agents #2 continued

- **Ingestion Exposure**
  - DO NOT induce vomiting

NOTE Systemic absorption can occur - closely monitor blood pressure, blood volume, liver and kidney function

ANTIDOTE

- Dimercaprol (BAL – British Anti-Lewisite)
  - BAL should be administered systemically when shock or significant pulmonary injury are present
  - 3 to 6 mg IM every 4 hours for 4 doses
  - Contraindicated in:
    - Pre-existing renal disease
    - Pregnancy (except in life-threatening circumstances)
    - Concurrent use of iron supplements

Blister Agents #2 continued

Laboratory Tests/Findings

- CBC
- Glucose
- Serum electrolytes
- Chest x-ray
- Arterial blood gases/pulse oximetry
- Urinalysis and thiodiglycol assay
- Urinary arsenic levels

Disposition/Follow-up/Reporting

- Admit patients with significant exposure with skin, eye, and airway symptoms
- Observe mildly symptomatic patients for at least 18 to 24 hours post-exposure
- Observe inpatients for delayed effects
- Instruct discharge patients to watch for delayed effects
- Follow-up laboratory evaluation for:
    - Pulmonary function
    - Renal function
    - Hepatic function
    - Bone marrow toxicity
- Notify state or local health department of possible public health risk

Choking Agents

Choking Agents

Phosgene

Chlorine

Contact Risk

- **High** – severe lung irritant, contact with vapor at high concentration can be fatal

Antidote

None

Routes of Exposure

- **Inhalation**
- **Skin/Eye contact**

Personal Protection

- See General Preparedness section for specific Personal Protection advice

Emergency Actions

- See General Preparedness section for specific Emergency Action advice

NOTE Low risk of secondary contamination from victims exposed only to gas

Signs and Symptoms

NOTE Children are more susceptible to effects of pulmonary agents

PHOSGENE:

- Irritation and pain of eyes, nose, mouth, pharynx
- Tearing, blood in eye
- Skin irritation, redness
- Cough, choking sensation
- Chest tightness
- Nausea, vomiting
- Headache
- Patient may by asymptomatic for 24 to 48 hours (latent period)**
- Progressive difficulty breathing (dyspnea)
- Severe, delayed-onset pulmonary edema (up to 24 to 48 hours post-exposure)
- Shock
- Death

** **NOTE** Serious pulmonary effects may not appear for 24 to 48 hours following exposure

CHLORINE:

- Irritation and pain of eyes, nose, mouth, pharynx
- Tearing, conjunctivitis
- Skin irritation, redness
- Cough, choking sensation
- Burning sensation in chest
- Rapid breathing, cyanosis
- Progressive difficulty breathing
- Pulmonary edema
- Death

Characteristics

Phosgene:

- Colorless gas at room temperature
- Odor (new mown hay)
- Volatile
- Vapors heavier than air
- Odor DOES NOT provide sufficient warning of hazardous concentrations
- May produce asphyxiation in poorly ventilated, low-lying or enclosed spaces
- Reacts with water to form hydrochloric acid
- Highly corrosive to lungs, eyes and skin
- Liquid agent may produce frostbite

Choking Agents continued

Characteristics continued

Chlorine:

- Yellow-green gas at room temperature
- Strong odor
- Volatile
- Vapors heavier than air
- Odor provides sufficient warning of hazardous concentrations
- Reacts w/water to form hydrochloric acid and hypochlorous acid
- Highly corrosive to lungs, eyes, skin
- May produce asphyxiation in poorly ventilated, low-lying or enclosed spaces
- Liquid agent may produce frostbite

Treatment: Pre-Hospital

HOT ZONE

(Use Level A PPE)

- Remove victims from incident site ASAP
- Decontaminate liquid agent from skin and clothing
- Execute ABC's on victims
- Triage patients (see Triage section)
- Check for obstructed airway – clear if necessary
- Move victims to Decontamination Site – for non-ambulatory, use gurney or backboard that can be decontaminated
- Keep victims warm and quiet: physical activity subsequent to exposure increases likelihood of death

NOTE Maintain records of all medications given

WARM ZONE (Decontamination Site)

(Use Level A PPE – lower level of personal protection permitted only if exposure levels are determined to be safe)

- Execute ABC's
- Keep victims warm and quiet
- Flush exposed eyes with water or saline for 15 minutes
- If exposure involved liquid agent – perform decontamination (see Decontamination section)
- If frostbite present, place frostbitten area in warm water (102 to 105 degrees Fahrenheit) until normal circulation returns
- No transfer to Cold Zone until completely decontaminated

NOTE All patients suspected of having phosgene exposure must be transported to hospital for evaluation and observation

COLD ZONE

- Verify decontamination
- Execute ABC's
- Observe for clear airway
- Intubate if necessary
- Administer supplemental oxygen
- Place on cardiac monitor
- Continue eye irrigation as indicated

Treatment: Pre-Hospital continued

- Continue frostbite rewarming as indicated
- Perform advanced treatment as necessary
  - Advanced life support protocols: coma, hypotension, cardiac arrythmias, and seizures
  - Treat bronchospasm with bronchodilators
  - Intubation, cricothyrotomy
  - Supportive care
- Transport to hospital
  - Contact and give them condition of patient, treatment already given, and Estimated Time of Arrival (ETA)

Treatment: Emergency Department

NOTE Contaminated patients may arrive at the ER without prior notification

DECONTAMINATION AREA

If exposure to liquid agent:

(Use Level A PPE for decontamination)

- Decontaminate prior to entry into facility
- Only decontaminate inside hospital if:
  - Use negative air pressure room
  - Use floor drains to contain contamination
- Execute ABC's

Choking Agents continued

TREATMENT AREA

- Verify decontamination
- Triage patients (see Triage section)
- Execute ABC's
- Enforce rest
- **Eye exposure**
  - Thorough eye examination
  - Visual acuity testing
  - Ophthalmology consult for corneal injuries
- **Skin exposure**
  - Thorough evaluation
  - Treat chemical burns and frostbite injuries
  - Small area: wound care, observation, follow-up
  - Large area: admit, wound care, observation, treatment (possibly in burn unit)

Treatment: Emergency Department continued

- **Airway exposure**
  - Thorough evaluation
  - Phosgene
    - Supportive care
    - Airway management/control secretions
    - Bronchodilators
    - Prevent/treat hypoxia
    - Observe for signs of respiratory distress up to 48 hours post exposure
    - Corticosteroids (following severe exposure)
    - Cardiotropic drugs (as indicated)
    - Fluid resuscitation
    - PEEP - positive end-expiratory pressure
  - Chlorine
    - Supportive care
    - Airway management/control secretions
    - Bronchodilators
    - Prevent/treat hypoxia
    - Observe for signs of respiratory distress
    - Corticosteroids (following severe exposure)
    - Cardiotropic drugs (as indicated)
    - Fluid resuscitation

Laboratory Tests/Findings

- CBC
- Glucose
- Serum electrolytes
- Chest x-ray
- Arterial blood gas measurements/pulse oximetry
- ECG

Disposition/Follow-up/Reporting

- Admit patients with significant exposure to skin, eye, and airway
- Admit patients with airway symptoms to ICU
- Admit all patients with suspected phosgene exposure – reexamine periodically – discharge after 48 hours if asymptomatic
- Observe inpatients for delayed effects – eyes, skin, lungs (RADS – Reactive Airway Dysfunction Syndrome)
- Instruct discharge patients to watch for delayed effects – eyes, skin, lungs (RADS)
- Notify state or local health department of possible public health risk

NUCLEAR TERRORISM

General Information

Personal Protection

RESPONDING TO AN EVENT

LEVEL A **Personal Protective Equipment (PPE)**

Provides highest level protection for respiratory, skin, eye, and mucous membrane danger

- Respiratory protection
  - NIOSH-approved positive-pressure, self-contained breathing apparatus (SCBA) if fire, smoke, fumes, chemicals, or heavy dust particles in the air (replace canisters and cartridges before using respirator)
  - Full face mask with HEPA filter*
- Radiation protective clothing
  - Nuclear, Biological and Chemical (NBC) suit – cover all exposed skin areas
  - Use a Toxic Agent Protective Apron (TAP Apron) for decontamination activities and patient handling
  - Double gloves
  - Protective boots

  NOTE Radiation protective clothing protects against alpha and beta particles, **NOT GAMMA RADIATION**
- Communication
  - Use two-way radio system

*Technically not LEVEL A, but acceptable in radiation incidents

NIOSH – National Institute of Occupational Safety & Health

REMOVAL OF PERSONAL PROTECTION

1. Remove protective clothing prior to leaving contaminated area
2. Disrobe head to toe (decreases likelihood of inhalation of radioactive material)
3. Shower and wash with soap and water - rinse

Characteristics

- Radiation cannot be seen, smelled, felt, or tasted
- Adverse health effects of radiation exposure may range from mild to serious
- Adverse health effects may not become apparent for many years
- Severity of effects dependent upon several factors:
    - Amount of radiation absorbed
    - Type of radiation
    - Route of exposure
    - Duration of exposure
- Exposure to very large doses may be fatal within a few days or months
- Exposure to lower doses may lead to increased risk of cancer and other chronic diseases

NOTE Health consequences of a nuclear terrorist attack will vary according to the type of attack and distance a person is from the attack

Emergency Actions

Control the Scene

- Approach from upwind with caution
- Determine safe site entry location
- Establish communications
- Implement nuclear response protocols
- Ensure safety of personnel
    - Thorough hazard assessment
    - Use PPE
    - Use buddy system
    - Monitor and respond to changing conditions
    - Do not eat, smoke, or drink in contaminated area
    - Avoid direct contact with radioactive materials/remote handling is recommended whenever possible (shovels, ropes, long-handled implements)
    - Minimize exposure time and contact
    - Rotate personnel as required
    - Monitor rescue personnel with individual radiation dosimeters, if available
- Control/extinguish fires
- Be aware of possible secondary devices

Emergency Actions continued

- Establish Site Control Zone (see chart below)
    - Determine levels of radiation/contamination
    - Establish incident control line
    - Limit access to authorized personnel
    - Reroute traffic and establish strict traffic control corridor into and out of contaminated area
    - Evacuate or shelter in place as indicated (see Important Issues for Survivors on page 104 or 110)
    - Insure radiation and contamination control
    - Cover detection equipment in plastic to minimize contamination

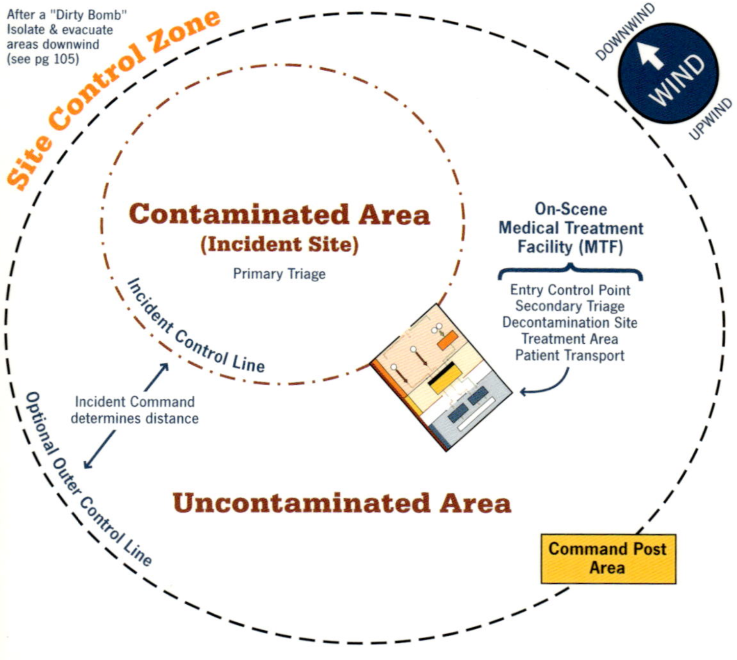

- Establish on-scene Medical Treatment Facility (see pgs 34 to 39)
    - Hot Zone – secondary triage, clothing removal
    - Warm Zone – Decontamination Site
    - Cold Zone – treatment and transport areas
- Establish and maintain crowd control
    - Prevent panic
    - Prevent public disorder

Rescue Victims

- Triage victims
- Assess and treat life threatening injuries immediately. Medical problems take priority over radiation concerns.
- Provide basic life support (ABC's)
- Evacuate victims/personnel to on-scene MTS
- Secondary triage for ambulatory patients
- Protect open wounds from radioactive contamination
- Disrobe victims – clothing removal is a major part of decontamination
- Decontaminate victims (full body rinse with lukewarm water)
    - **In a Contained Casualty Situation**
      water rinse, soap/water wash, water rinse
    - **In a Mass Casualty Situation**
      Low pressure water spray (fire hose, portable showers, mobile decontamination units, fixed facilities away from incident site)
    - After clothing is removed and skin decontaminated, victims are not radioactive and pose no danger to first responders
    - Redress patients in clean clothing

Emergency Actions continued

- Provide Advanced Trauma Life Support (ATLS)
- Notify treatment facilities, local, state agencies
- Transport victims (see Patient Transport on the next page)

Additional Response Actions

- Evaluate rescuers and equipment post-entry
- Conduct thorough site evaluation
- Preserve incident site as crime scene
- Collect evidence
- Establish/conduct proper record keeping
    - Names and addresses of all persons involved
    - Make detailed records of the incident
- Wrap, label, and isolate all equipment, clothing, tools, etc. used in the controlled area and retain until cleared by radiation safety authority
- Accomplish site isolation and clean-up
- Remove protective gear/clothing at incident control line (see page 92 for removal of personal protection)
- Delay cleanup pending instructions from radiation authority
- Radiation detection instruments necessary to:
    - Measure background radiation
    - Verify decontamination
- Levels of contamination (2x background = contamination)

Patient Transport

- Coordinate with overall response plan
- Verify decontamination prior to transferring patients
- Transport victims to designated hospital
- No transport without prior notification
  - Number of patients
  - Types and severity of injuries
- Use ground transport, NOT AIR
- Identify decontaminated patients clearly

NOTE Contaminated patients may arrive at treatment facilities without prior notification

Hospital Emergency Management

- Review protocols for evaluation, decontamination and treatment
- Assemble radiological emergency response team
- Breakout PPE and decontamination supplies
  - Surgical scrub suit, gown, mask, cap, double gloves, eye protection
  - Waterproof shoe covers
  - Tape all cuffs and open seams
  - Radiation dosimeter attached to outside of clothing
  - Waterproof apron
- Perform pre-arrival determinations
  - Number of victims
  - Medical condition of individual victims
  - Treatment rendered to victims / decontamination status
  - Estimated Time of Arrival (ETA)

Hospital Emergency Management continued

- Notify local emergency authorities
- Establish security & traffic control of hospital entrances & grounds
- Clear and secure all areas to prevent contamination
- Establish methods for contamination control
- Prevent entry of victims into emergency department or MTF without adequate evaluation and decontamination
- Establish controlled area for patient arrival, decontamination, treatment and holding (Hot, Warm, Vapor Hazard & Cold Zones)
- Establish control line between controlled & uncontrolled areas
- Secure and control traffic to and from controlled area
- Establish secondary control line
- Prevent tracking of contamination by covering floor areas (brown wrapping paper, butcher paper, cloth sheets, taped securely to floor)
- Establish strict isolation precautions
- Control waste disposal and ventilation
- Check radiation survey instruments and determine background radiation levels
- Decontaminate patients before transfer out of controlled area
- Verify patient and personnel decontamination prior to exit from controlled area
- Transfer patients only with prior notification
- Establish accurate record keeping for patient management and identification

Acute Radiation Syndrome (ARS)

Result of high doses of penetrating radiation to much of body, over a very short period of time (usually minutes)

IMMEDIATE SYMPTOMS

- Nausea
- Vomiting
- Diarrhea
- Skin damage / redness
- Hair loss
- Asymptomatic period (2 to 3 weeks)

SECOND-STAGE SYMPTOMS

- Bone marrow depletion (2 to 3 weeks)
- Loss of appetite
- Hair loss
- Infection
- Bleeding
- Malaise
- Possible death
- Survival dependent upon radiation dose

Treatment

- Supportive therapy only
- Recovery process may last weeks and up to 2 years

"Dirty Bomb"

"Dirty Bomb"

Radiological Dispersal Devices (RDD)

Contact Risk

- High

Antidote

- None

Definition / Characteristics

- Improvised explosive devices combining explosives with radioactive materials
- Non-explosive devices used to spread radioactive material (solid, liquid, powders) spread covertly
- Explosive blast is primary danger
- Serious injuries and fatalities possible
- Ineffective for exposing large numbers of people to lethal doses of radiation
- Generally affects only a limited area

Purpose of a "Dirty Bomb"

- Incite fear and panic among people
- Disperse radioactive materials
- Make facilities and land unusable

"Dirty Bomb" continued

Routes of Exposure

(can happen in combination)

- External radiation (absorbed by or passes through body)
- Contamination (external and internal)
    - Liquid
    - Solids
    - Gases
- Incorporation (radioactive material taken up by cells, tissues, and organs of body)

Personal Protection

- See General Information section for specific Personal Protection advice

Emergency Actions

- See General Information section for specific Emergency Action advice
- Take action to minimize further release of radiation (if possible)
- Define areas with highest radiation levels
- Do not enter highest dose rate areas except to save lives, make entry as brief as possible

IMPORTANT ISSUES FOR SURVIVORS

1. Evacuate, DO NOT SHELTER IN PLACE
2. Decontamination

Critical Considerations

- Primary types of injuries:
  - Blast injuries (internal organs, eardrums)
  - Thermal burns
  - Multiple trauma
  - Radiation exposure / contamination

NOTE DO NOT let radiological contamination interfere with immediate life-saving treatment of best possible medical care

- After clothing is removed and skin decontaminated, victims are not radioactive and pose no danger to first responders
- Use of KI tablets (potassium iodide) **IS NOT** recommended following detonation of radiological dispersion devices

Initial Isolation & Evacuation Distances

ISOLATE & EVACUATE THE AREA DOWNWIND FROM INCIDENT SITE

Isolation Distances		Small Incident	Large Incident
	(feet)	150	300–600
Evacuation Distances		**No Fire**	**Fire**
	(feet)	1000	2640

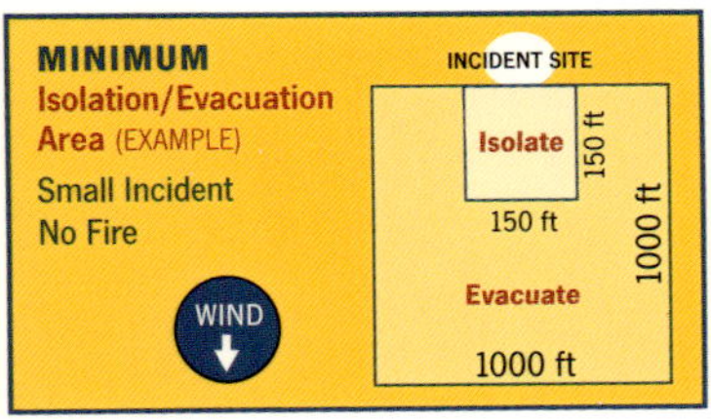

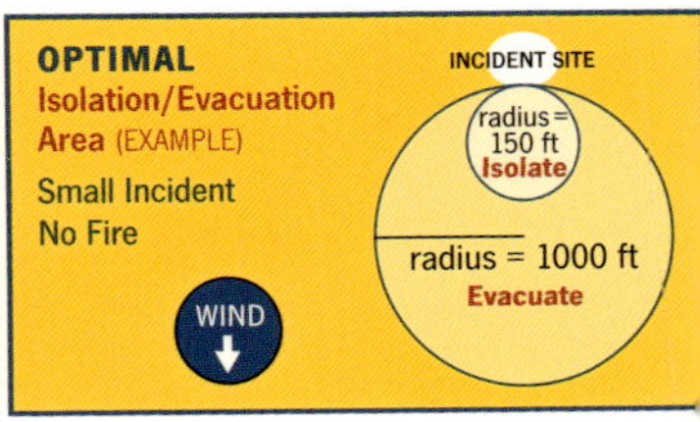

"Dirty Bomb" continued

Further Evacuation Information

- Ultimate area requiring permanent evacuation may be much larger than original incident site
- Dispersal of radioactive contamination dependent upon
  - Size and sophistication of weapon
  - Wind and atmospheric conditions

Patient Transport

- See General Information section for specific Patient Transport advice

Hospital Emergency Management

- See General Information section for specific Hospital Emergency Management advice

Nuclear Detonation

Nuclear Detonation

Contact Risk

- **High**

Antidote

None

Routes of Exposure

(can happen in combination)

- External radiation (absorbed by or passes through body)
- Contamination (external and internal)
    - Liquid
    - Solids
    - Gases
- Incorporation (radioactive material taken up by cells, tissues, and organs of body)

Characteristics

- **See General Information section for general Characteristics information**
- Catastrophic event
    - Massive destruction
    - Thousands of people:
        - Dead
        - Injured and contaminated
        - Asymptomatic but with serious radiation exposure
    - Nuclear fallout, nuclear cloud (effects dependent on atmospheric conditions, wind speed and direction)
    - Damage to structures, infrastructure

Characteristics continued

- Catastrophic event continued
  - Disruption of:
    - Civil authority
    - Communications
    - Transport
    - Response capability
- Blast effect
- Thermal radiation
- Ionizing radiation
- Electro-magnetic pulse

Personal Protection

- See General Information section for specific Personal Protection advice

Emergency Actions

- See General Information section for specific Emergency Action advice
- Stay away from ground zero and enter surrounding areas only to save lives (radiation levels will be very high)

IMPORTANT ISSUES FOR SURVIVORS

1. Shelter in place for at least the first 48 hours
2. Decontamination

Critical Considerations

- Following a nuclear detonation, transport, communications and capabilities of medical treatment facilities may be non-functional or overwhelmed.
- Burns and traumatic injuries are the most immediate consequences of nuclear detonation and will require conventional medical and surgical treatment
- Physical injuries will be more serious than radioactive contamination
  - Patient stabilization is first priority
  - Patient decontamination is second priority

NOTE **DO NOT** let radiological contamination interfere with immediate life-saving treatment of best possible medical care

- After clothing is removed and skin decontaminated, victims are not radioactive and pose no danger to first responders
- Expect significant numbers of patients to develop Acute Radiation Syndrome as a result of exposure at time of detonation and exposure to fallout
  - See General Information section for specific ARS treatment advice

Patient Transport

- See General Information section for specific Patient Transport advice

Hospital Emergency Management

- See General Information section for specific Hospital Emergency Management advice